W9-ATT-219

Pat Schories

OVER UNDER IN THE GARDEN

An Alphabet Book by Pat Schories

Farrar, Straus and Giroux · New York

Published simultaneously in Canada by HarperCollins*CanadaLtd*
Color separations by Hong Kong Scanner Arts
Printed and bound in the United States of America by Worzalla
First edition, 1996

Library of Congress Cataloging-in-Publication Data. Schories, Pat. Over under in the garden : an alphabet book / Pat Schories. p. cm. 1. Plants, Ornamental—Juvenile literature. 2. Vegetables—Juvenile literature. 3. Garden fauna—Juvenile literature. 4. English language—Alphabet—Juvenile literature. [1. Flowers. 2. Vegetables. 3. Gardens. 4. Garden animals. 5. Alphabet.] I. Title.

SB406.5.S34 1996 635.9—dc20 95-275 CIP AC

ACORN

Ant

BLACK-EYED SUSAN *Buckeye butterfly*

CATNIP

Caterpillar

DOGWOOD

Daddy Longlegs

EGGPLANT

Elephant stag beetle

FOXGLOVE *Firefly*

GRAPE

Grasshopper

HONEYSUCKLE *Hornet*

IRIS

Inchworm

J

JACK-IN-THE-PULPIT

Japanese beetle

KOHLRABI *Katydid*

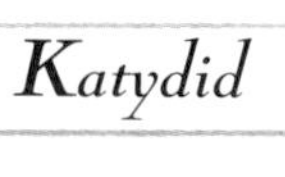

L

LILY PAD

Ladybug

MANDRAKE

Mosquito

N

NASTURTIUM

Net-winged beetle

OBEDIENT PLANT

Orb weaver

PEA

Praying mantis

Q

QUEEN ANNE'S LACE

Queen butterfly

ROSE

Ruby spot

S

STRAWBERRY *Stinkbug*

T

TOMATO

Tiger beetle

UMBRELLA PLANT

Underwing moth

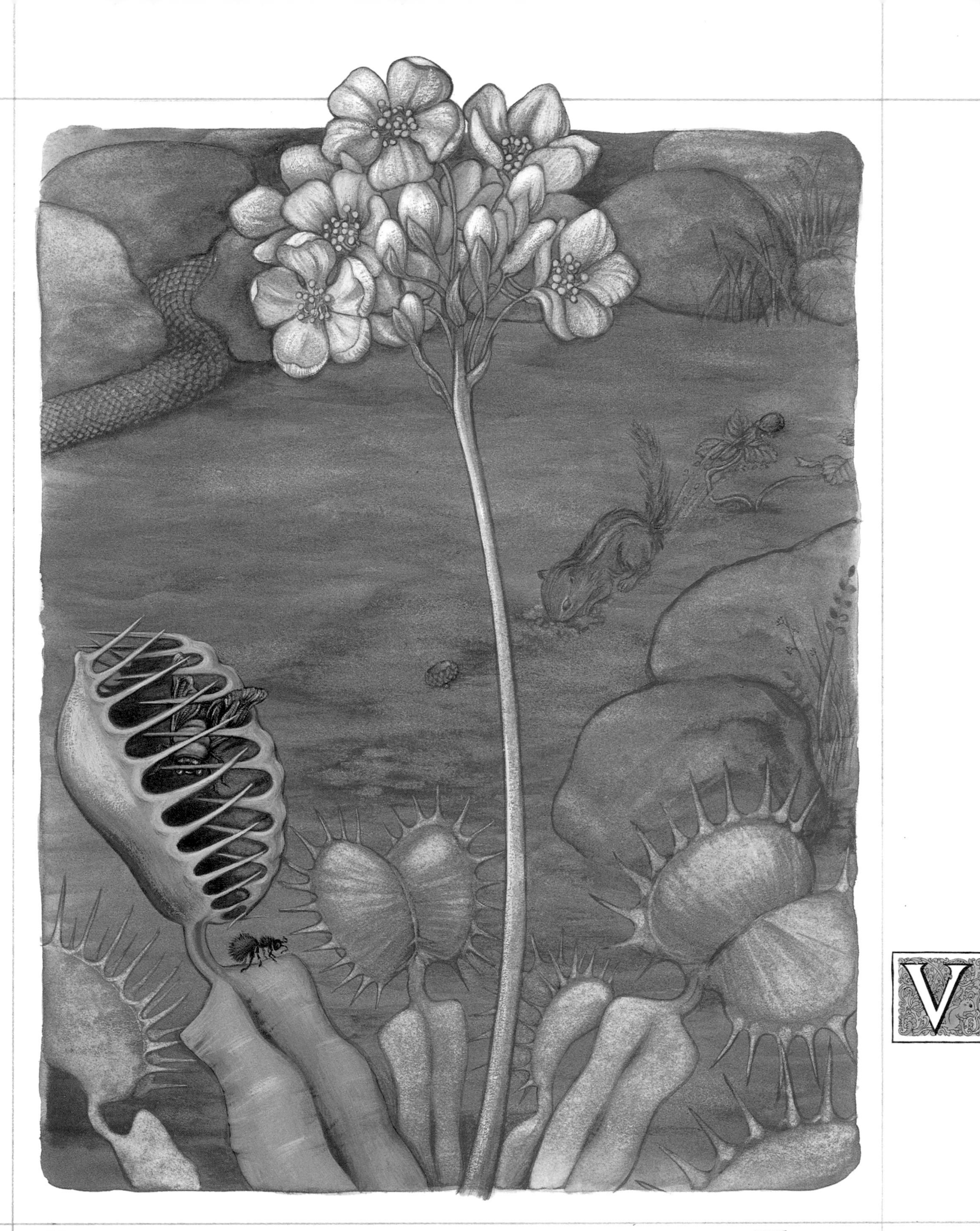

VENUS FLYTRAP

Velvet ant

WISTERIA

Walking stick

XERANTHEMUM

Xenopsylla cheopsis (flea)

Yucca

Yellow jacket

Zucchini

Zebra butterfly